# Rebuilding Trust Between Silicon Valley and Washington

# COUNCIL *on* FOREIGN RELATIONS

Council Special Report No. 78
January 2017

Adam Segal

# Rebuilding Trust Between Silicon Valley and Washington

# Contents

# Foreword

The global cyberspace landscape is best understood as a modern Wild West, with many gunmen, few laws, and no sheriff. Not surprisingly, cybersecurity has emerged in recent years as one of the most consequential and controversial realms of foreign policy and international relations. From the pilfering of enormous amounts of what was private data from the U.S. Office of Personnel Management, to the theft of customer information from Target, JPMorgan Chase, and numerous other corporations, to North Korea's 2014 attack on Sony Pictures, to the 2016 hacking of the Democratic National Committee and others, the number and frequency of cyberattacks in and against the United States—including its government, corporations, and citizens—is growing.

At the same time, other cyber issues are emerging, including debates about international jurisdiction over data, which have led several countries to localize data in their own territories, and over data encryption, which enhances privacy but leads to questions about security. Also more prevalent are actions by states to restrict internet access and capabilities for their own populations. And there is the reality of and potential for using cyber tools not just for espionage but for an act of sabotage and war.

The divisions and differences between the U.S. government and the American technology community have also grown. The National Security Agency revelations from Edward Snowden and policy disagreements on encryption and data accessibility, among other factors, have led to a feeling of mutual distrust between the public and private sectors. The government tends to emphasize matters of national security; corporations tend to most value consumer preferences, fearing they will forfeit their market position if they are seen as getting too close to authorities. This divide has led to U.S. policymaking that is ill equipped to keep up with technological advancements and changes in the cyber environment. It has also complicated the effort against terrorism and stymied the United States' ability to work with allies abroad to generate consensus on cyber norms.

In this Council Special Report, Adam Segal, the Ira A. Lipman chair in emerging technologies and national security and director of the Digital and Cyberspace Policy program at the Council on Foreign Relations, offers several policy areas where Washington and Silicon Valley can and should work together. These include creating a devoted, advanced cyber workforce for the U.S. government, combating data localization trends, and deterring state actors in a way fit for the global cyber era. Most difficult, he writes, will be collaborating to establish norms suitable to both constituencies on data encryption and access. Segal offers some concrete recommendations for the government and technology community to take in order to create real advancements in these realms, such as expanding existing programs that bring high-skilled workers to the government for short projects, attributing attacks and responding with options such as sanctions, providing more clarity on the U.S. judicial process for foreign governments and companies, and allowing "lawful hacking" under certain circumstances with strict oversight.

The issues faced in the cybersecurity realm are and will remain numerous. For the United States to reduce its vulnerability to economic, strategic, and political cyberattacks—and for the U.S. technology industry to continue to thrive globally—it is important that the two constituencies find ways to work together. Both groups would be wise to consider Segal's thoughtful and practical recommendations when shaping their relationship in the coming months and years.

**Richard N. Haass**
*President*
Council on Foreign Relations
January 2017

# Acknowledgments

I would like to express my gratitude to the many people who made this report possible. To begin, I would like to thank CFR President Richard N. Haass and Director of Studies James M. Lindsay for their support of this project and insightful feedback throughout the drafting process.

I would like to thank the members of the CFR cyber standing working group, which met in Washington, DC, and Silicon Valley, for their ideas, expertise, time, and support. In particular, I would like to thank Craig James Mundie for chairing the group and leading the discussions as the group met through 2015 and 2016. Robert O. Boorstin, David P. Fidler, Tressa Guenov, Catherine B. Lotrionte, Jeff Moss, and Neal A. Pollard went above and beyond the call of duty, providing written comments that sharpened the report's arguments.

I am grateful for the valuable assistance of Patricia Dorff, Elizabeth Dana, and Erik Crouch in CFR's Publications Department, who provided unmatched editing support, and to Melinda Wuellner and Andrew Palladino in Global Communications and Media Relations for their outstanding marketing efforts. I also appreciate the contributions of the David Rockefeller Studies Program staff, including Amy Baker, in shepherding the report.

Tremendous thanks go to the members of the CFR Digital and Cyberspace Policy program, especially Assistant Director Alex Grigsby. The report would not have been completed without his help. I am also thankful for the assistance of Lincoln Davidson, who was then a research associate in the program.

This publication is a product of the Digital and Cyberspace Policy program. The meetings of the Council on Foreign Relations Working Group on Cyberspace and U.S. Foreign Policy were supported in part by funding from PricewaterhouseCoopers. CFR expresses its thanks to the Hewlett Foundation and LinkedIn for hosting several meetings of the working group at their offices in California, and to Ira A. Lipman for his support for the study of emerging technologies on U.S. national security. The statements made and views expressed herein are solely my own.

Adam Segal

*Council Special Report*

# Introduction

Despite significant new executive action, legislation, funding, institutional innovation, and diplomatic agreements over the last eight years, the threat of cyberattacks to U.S. economic, strategic, and political interests continues to grow. Although the Barack Obama administration signaled early that it intended to make cybersecurity a priority, the strategic situation has not significantly improved, despite the White House's efforts. Cybercrime is growing in volume and sophistication, and some governments have become more brazen in using cyber operations for espionage, coercion, and influence. The vast majority of these incidents are disruptive, designed to undermine trust in complex economic, political, and social institutions. Meanwhile, new vulnerabilities are developing with the emergence of the internet of things: millions of devices fitted with sensors that collect data and communicate over the internet. A future generation of attacks on the internet of things could cause widespread economic dislocation and physical destruction.

In addition, the free flow and physical location of digital data have become significant sources of conflict in U.S. trade and foreign policy. Motivated by law enforcement and economic concerns, many countries are passing laws that require user data be stored within their borders, a trend known as data localization. Because U.S. companies are legally prohibited from releasing content data to foreign governments without a warrant, these governments are pushing to keep data inside their jurisdictions. Some foreign governments also believe that storing data locally can help domestic technology firms and spur local innovation. With populist and antiglobalization sentiment growing in major economies around the world, data nationalism may become an even more pronounced force, and U.S. technology companies may find themselves faced with having to choose between competing authorities and jurisdictional conflicts.

The United States has historically addressed cybersecurity challenges with an incremental approach, based on public-private partnerships, information sharing, international partnerships, norms development, and deterrence. However, these policymaking models have not kept pace with rapid technological change and the proliferation of malicious actors and threats. In addition, many lawyers, judges, legislators, and decision-makers lack the technical skills to comprehend the software, hardware, and networks that make up cyberspace.

More agile, flexible, and creative policymaking is required, but effective cyberspace policy cannot be crafted without the cooperation of the private sector. Private companies own the majority of the internet's infrastructure and are the major drivers of technological innovation. The Donald J. Trump administration will have to confront the deep mistrust between Washington and the information technology community—much of which resulted from the disclosures by National Security Agency (NSA) contractor Edward Snowden—as well as arguments over encryption and lawful access to data. Some of the documents Snowden gave to the press showed not only that technology companies were legally required to turn over the data of non-U.S. citizens to the government, but also alleged that the NSA was inserting malware into companies' products, undermining the standards that are the basis of encryption widely used in commercial products, and hacking into several companies' cables carrying the data among servers in different parts of the world. All of these actions damaged the reputation of U.S. technology companies with their users and catalyzed foreign governments to exert more control over transnational data flows.

If the United States does not address the growing divide between the government and the technology community, it will lose influence in cyberspace and be unable to keep pace with cybersecurity threats. Although numerous government officials have traveled to Silicon Valley over the past several years, narrowing the gap will not be easy, in part because technology firms operating as global platforms have strong economic motivations to keep Washington at a distance. Potential adversaries will continue to use hardware and software developed by U.S. companies, and thus law enforcement and intelligence agencies will persist in exploiting the vulnerabilities in these products. Still, there are four policy areas where meaningful progress can be made, ranging from relatively easy fixes to more difficult challenges. The private sector and the U.S. government have a shared interest in, first, creating a vibrant cyber workforce and, second, fighting the global trend of forced data localization. The third area, deterring state attackers, is necessary but more difficult and will demand new conceptual models that rely less on Cold War history. A workable compromise over the deployment and use of encryption and lawful access to data, the fourth issue, would be the most consequential step in restoring trust, but also the most demanding.

In order to make progress in these four areas, the United States should

- continue support for the U.S. Digital Service (USDS) and create a highly specialized cybersecurity service within the U.S. government;
- amend provisions of the Electronic Communications Privacy Act, using the U.S.-UK agreement as a template, to allow technology companies to provide data to foreign governments that meet specific requirements;
- attribute attacks more frequently and, for cyberattacks that fall below the use of force and armed attack threshold, devise and implement forceful responses, such as covert cyber operations designed to disrupt future attacks; and
- strengthen law enforcement's ability to conduct lawful hacking under strict judicial oversight and a clearly defined vulnerabilities equity process.

These recommendations are no panacea to the lack of trust between Silicon Valley and Washington. In February 2016, the Federal Bureau of Investigation (FBI) successfully obtained a court order requiring Apple to build software to unlock the iPhone of one of the San Bernardino terrorists, generating a sharp divide between the technology community and law enforcement over the legitimate uses of encryption. As a candidate, Donald J. Trump targeted several tech companies for sending jobs overseas. Many in the technology community actively supported Democratic nominee Hillary Clinton and had close ties to the Obama administration. Differences over immigration, trade, climate change, net neutrality, and antitrust regulation are stark. Although there is little hope that trust between the two sides will be reset to a pre-Snowden level, progress in these four areas will at least prevent a further deterioration of relations that would worsen U.S. cybersecurity. At best, addressing these areas will provide a new, pragmatic basis for industry-government cooperation.

# Growing Threats

The United States faces cybersecurity risks that threaten its economic, political, and strategic interests. Cybercrime continues to grow in sophistication and magnitude, costing the global economy an estimated $500 billion annually.[1] Ransomware attacks—the encryption of data until a ransom is paid—rose to an average of four thousand attacks per day in the first quarter of 2016, four times more than the daily average in 2015.[2] In 2015, the FBI estimated that victims paid about $40 million in ransoms; that number may rise to $400 million in 2016. Hackers breached 113 million health care records in 2015, up from 12.5 million in 2014.[3]

Foreign governments have grown increasingly able and willing to use cyberattacks for sabotage, espionage, and political influence. Russian hackers are suspected of being behind an attack on a Ukrainian power grid in December 2015 that left 230,000 residents without power for several hours. Russian hackers are also reportedly responsible for the theft and public release of documents from the Democratic National Committee and the Democratic Congressional Campaign Committee in the summer of 2016, and they may have also been behind compromises of local election systems in Florida, Arizona, and Illinois.[4] In 2014, Chinese hackers allegedly stole personal data of more than twenty million U.S. government workers from the Office of Personnel Management.[5] Iranian hackers were reportedly responsible for widespread distributed denial of service (DDoS) attacks on U.S. financial institutions, whose websites were flooded with junk traffic, bringing them down; an attack on Saudi Aramco in 2012 that forced the oil giant to replace thirty thousand computers and take its business networks offline for two weeks; and the breach of computers at a dam in Rye, New York. In December 2014, North Korean hackers breached Sony Pictures' computers, knocking them offline and causing millions of dollars in damage in retaliation for a movie insulting the nation's dictator.[6]

The majority of attacks are designed for espionage, disruption, or political influence; they do not cross into the territory of an armed attack or the use of force. The nature of the threat, however, will change as the internet of things—robots, cars, medical equipment, and other machines that communicate over the internet—introduces new vulnerabilities. Gartner, a technology research firm, estimates the number of devices connected to the internet will rise from 6.4 billion in 2016 to 20 billion by 2020.[7] In October 2016, tens of thousands of internet-connected cameras made by Chinese companies were commandeered to attack Dynamic Network Services in one of the largest DDoS attacks ever recorded, resulting in the disruption of some of the internet's largest sites, including Netflix, Twitter, PayPal, and Reddit. The former Director of National Intelligence James Clapper has warned that "devices, designed and fielded with minimal security requirements and testing, and an ever-increasing complexity of networks could lead to widespread vulnerabilities in civilian infrastructures and U.S. government systems."[8]

The international politics of cybersecurity and data flows are also becoming increasingly contentious. Nations have reasserted control over their citizens' information through data localization, censorship of information, and other forms of so-called cyber sovereignty. Traditional conceptions of territoriality and sovereignty have not kept up with global platforms, and legal definitions, public policy goals, and social expectations of privacy are diverging between the United States, Europe, and Asia. This scenario creates an unpredictable and potentially costly environment for global corporations whose data does not recognize borders. Moreover, antitrade sentiment and rising nationalism may increasingly have a digital component, further challenging the view of the internet as an open, global platform.

In addition to rising cybersecurity threats, the Trump administration will inherit a growing political divide between Washington and U.S. tech firms that stems in large part from the disclosures by NSA contractor Edward Snowden. Angered by reports that the NSA allegedly undermined encryption, placed back doors into products, and hacked into cables carrying data traffic, companies and human rights advocates alike have legally challenged the White House and have supported the introduction of new technologies designed to hamper surveillance. Encryption has become a particularly contentious issue in the relationship between the U.S. government and domestic tech firms; law enforcement agencies have warned that they are unable to access encrypted evidence even with legal authority, a problem known as going dark. Meanwhile, the technical community argues that giving law enforcement access to encrypted data to identify a few bad actors will weaken cybersecurity and privacy for everyone.[9] Google, Apple, Yahoo, Microsoft, and others have fought back by rolling out encryption for a growing number of services and products, lobbying for intelligence reform, and challenging the government in court over access to customer emails and other personal data stored outside the United States and for the right to notify customers when the U.S. government asks for user information.[10] With their dependence on foreign markets for revenue and growth increasing, U.S. technology firms have often aligned themselves rhetorically with those skeptical of U.S. intelligence practices who prioritize security and privacy of all internet users, in contrast to the U.S. government's focus on national security. During his campaign, Donald J. Trump called for a boycott of Apple products and sided with the FBI during its court battle with the company, exacerbating tensions between Washington and Silicon Valley.[11]

Technology companies do not speak with one voice, and their economic interests differ, depending on their size, sector, and location. Some companies go public with their differences with Washington; others continue to cooperate quietly and still others do both, depending on the issue. But no matter how vocal they are in their dissent, all firms that generate, analyze, or store data will need regulatory consistency. They require consistent guidelines on how to respond to demands for access to user information both at home and abroad, and a clear sense of how the government is shaping cybersecurity.

After many years on the margins, cybersecurity is now a major area of policymaking. Yet the U.S. government, technology companies, privacy nonprofits and activists, and others involved in the policy debate have often delayed making difficult decisions. There has been a hope that hard work, a willingness to listen, and technological ingenuity would result in outcomes that avoided inevitable tradeoffs in core U.S. values such as security, privacy, accountability, and transparency. This delay has political costs, especially on the diplomatic front; the United States cannot forge a consensus with its friends and allies when it has failed to do so at home. It also has economic costs: as the technologies connected to the internet become more diverse, it becomes increasingly more difficult and expensive to retrofit complex systems to the demands of policymakers.

The Trump administration faces many challenges: rising cyber-crime and increasingly sophisticated and aggressive state-sponsored attackers, a growing divide with the technology community, and the reassertion of national sovereignty in cyberspace. These problems are interconnected and will require sustained focus from policymakers as well as high levels of cooperation with the private sector.

# Cyber Policy Principles and Progress

When President Barack Obama took office in January 2009, the guiding principles of cybersecurity policy were already agreed upon and had broad bipartisan support. Under Presidents Bill Clinton and George W. Bush, U.S. domestic cyber strategy consistently stressed the need for private sector leadership and the importance of developing public-private partnerships and information sharing. Abroad, the administrations focused on creating partnerships, developing norms of state behavior, and strengthening efforts to deter potential adversaries.

However, this broad consensus on the direction of policy has rarely translated into efficient, timely policymaking, especially in Congress, due to factors such as the rapid pace of technological change, a lack of technical expertise among lawmakers, and political divisions over individual privacy and the role of the government.

Facing these challenges, the Obama administration nonetheless made a remarkable amount of progress reforming and implementing cybersecurity policy and strategy. Over the course of eight years, the Obama administration appointed the country's first chief information security officer, gave the executive branch the power to sanction individuals conducting malicious cyber activity, promoted the sharing of threats and vulnerabilities between private sector entities, explicitly recognized the importance of cyber capabilities to support military objectives in official doctrine, and ordered U.S. Cyber Command to disrupt computer networks used by the self-declared Islamic State.[12]

On the diplomatic front, the United States released its first international strategy for cyberspace and can now claim two successes. First, after years of accusing China of cyber espionage for commercial gain, Washington reached an agreement with Beijing that "neither country's government will conduct or knowingly support cyber-enabled theft of intellectual property, including trade secrets or other confidential business information, with the intent of providing competitive advantages to companies or commercial sectors."[13] As of November 2016, the agreement appeared to have contributed to a decline in the number of incidents attributed to China.[14] Although some experts remain skeptical about the agreement, the deal provides some basis by which the United States can hold China to account for its espionage activity for commercial gain.[15]

Second, a group of government experts at the United Nations, which included representatives from China, Russia, the United States, and other countries, agreed to U.S.-championed norms stating that international law applies in cyberspace and that states should not conduct activity that intentionally damages critical infrastructure or interferes with another country's cyber emergency responders.[16] It remains to be seen whether these principles will be upheld during a military or diplomatic crisis between states—say, an incident in the South China Sea that involves U.S. and Chinese forces.

Private sector attention to cybersecurity has also grown significantly in recent years, after formidable attacks on major companies including Target, Home Depot, Anthem, Sony, and JPMorgan Chase shook customer confidence. After suffering an attack in 2014 that compromised more than 83 million accounts, JPMorgan Chase doubled its annual budget for cybersecurity, bringing it to $500 million.[17] And in a 2016 survey of nearly six hundred corporate chief information officers, 46 percent reported a heavy focus on cybersecurity, up from only 13 percent in 2009.[18] Analysts project that global spending on cybersecurity will grow between 8 percent and 15 percent annually over the next five years.[19]

Increased investment, focus, and training in cybersecurity, however, should not distract the new administration or the private sector from the need for sweeping change. The rising influence of state-backed cyberattacks shows the problems with incremental reform of cyber policy. Previous strategies for cybersecurity—such as adjusting and enhancing existing methods for sharing information, developing international norms, and facilitating public and private partnerships—will not keep pace with the threat. The Trump administration should embrace sweeping change, not continue a status quo of incremental adjustments.

# Cyber Workforce

Demand for cybersecurity skills has risen dramatically in the past five years, outpacing supply. In 2015, more than two hundred thousand cybersecurity jobs went unfilled in the United States.[20]

The U.S. government will always find it challenging to recruit cybersecurity professionals. The higher pay packages in the private sector pose an obvious barrier to government recruitment, but there are also nonfinancial factors. Many entry-level cybersecurity professionals perceive the workplace culture at the Department of Homeland Security, FBI, and other government agencies as incompatible with their cultural preferences. More senior and experienced talent find the security background checks, financial reporting, and other bureaucratic hurdles too demanding and costly.

For the private sector, there is not only a shortage of qualified job candidates but also a gap between employer expectations and employee skills. Few computer science graduates are required to take a security or risk management class as part of their degrees.[21] Moreover, a university-level education may not prepare someone for a career in cybersecurity, especially because cybersecurity skills are often vocational in nature and learned on the job.

Existing efforts to develop more appropriate cybersecurity curricula and diversify the workforce to include more women and minorities are important to filling the skills gap. There is a real risk, however, that automation, machine learning, and other technological innovations could reduce the number of available cybersecurity jobs, leaving large cohorts of unemployed graduates.

A highly specialized cybersecurity service could help the government address the most pressing challenges. After the failed rollout of Healthcare.gov, the health insurance exchange website operated by the federal government, the Obama administration in 2014 created the U.S. Digital Service, a technology consulting team drawn from the private sector. Individuals spend as short as three months with the program and have been involved in projects such as developing apps for the Department of Veterans Affairs; digitizing applications and review processes for U.S. Citizenship and Immigration Services; and launching a bug bounty program for the Pentagon, which allows vetted computer security experts outside government to identify software flaws in Department of Defense systems in return for monetary reward.[22]

Beyond bringing new approaches to bureaucratic reform, these programs have a wider, symbolic effect. Schemes like the USDS not only allow the government to take advantage of the desire for public service that many technology entrepreneurs are looking to pursue but also are interpreted by the same group as a sign that the administration is "tech savvy," willing to embrace start-up culture, and optimistic about the impact of technology on society, the economy, and politics.[23]

Congress is considering legislation that would extend the operations of USDS until at least 2026.[24] The Trump administration should also create a time-limited fellowship program specifically focused on cybersecurity, modeled on the Epidemic Intelligence Service (EIS) at the Center for Disease Control and Prevention (CDC), a two-year training program focused on fieldwork. Fellows in a cyber intelligence service would serve for three years and be deployed to areas of the federal government where breaches have occurred and be tasked to fix and implement corrective measures to prevent future breaches. The specialized service would develop important in-house expertise and esprit de corps that could help with recruitment and, even more important, retention of talent within the government.

# Data Localization

Forging a joint framework to respond to growing international demands for access to data would also narrow the divide between Washington and U.S. technology companies. Numerous countries have passed or are considering regulations that would require user data be stored locally, and U.S. technology companies often find themselves forced to choose whether to ignore certain local laws. India, Indonesia, Malaysia, Nigeria, South Korea, and Vietnam have all adopted data localization provisions, and foreign businesses are protesting similar regulations in China's cybersecurity law.[25] In July 2016, Russian President Vladimir Putin signed a law that requires telecommunications and internet companies to retain copies of all the contents of communications for six months and store the data inside Russian territory. A few months later, Roskomnadzor, Russia's communications authority, announced that it would block LinkedIn after the social networking site was found to have violated the data storage law.[26]

The push to keep data within national borders has been driven by various countries' law enforcement agencies and also by widespread frustration with the time-consuming and confusing legal processes involved in acquiring data from other nations. The Electronic Communication Privacy Act prohibits U.S. companies from releasing users' communications to foreign governments or authorities without a warrant from a U.S. judge, a process that is long and arduous.[27]

If a Brazilian citizen, for example, uses a Microsoft messaging app to plan and execute a bank robbery in Rio with other Brazilian citizens, Microsoft cannot disclose the messages directly to the Brazilian police. Instead, Brazil has to request assistance from the U.S. Department of Justice to petition a U.S. judge to obtain the communications on behalf of Brazil. Known as the mutual legal assistance treaty (MLAT), this process can be opaque, time-consuming, and challenging for foreigners unfamiliar with the U.S. justice system. An MLAT request generally takes ten months to process, and as more communications are stored online, this process will become even more cumbersome during the Trump administration.[28]

There are also economic and technological motivations for countries' data localization requirements. Many politicians abroad believe that data localization policies will spur indigenous technological innovation and create jobs, despite evidence to the contrary. For instance, a 2016 study found that full implementation of all data localization and associated regulations in the European Union would lead to a 0.48 percent decline in real gross domestic product (GDP).[29] Moreover, following the Snowden revelations, many countries distrust U.S. tech firms and want to protect users from U.S. surveillance. There is little assurance that localized data would provide greater security, however, because U.S. intelligence agencies face fewer legal constraints when trying to access non-U.S. citizens' data abroad than they do when it is stored in the United States.

The United States and United Kingdom negotiated an agreement that would allow UK law enforcement agencies to request stored data and live intercepts directly from U.S. service providers, as long as the warrants did not target U.S. persons (defined to include U.S. citizens, legal permanent residents, and anyone physically present in the United States).[30] The Obama administration requested that Congress enact the legislative changes required to implement this agreement and to allow future administrations to sign similar deals with other countries.[31] This would allow U.S. service providers to comply with foreign investigations when the requesting government has a legitimate interest in the criminal activity being investigated and they meet the following criteria: the target is not a U.S. person, the request is subject to judicial oversight by the country making the request, the requesting government complies with international human rights and rule of law standards, and the request is narrowly defined and limited to the particulars of a specific investigation (i.e., bulk collection is prohibited).[32] The Trump administration should continue this effort and work with Congress to ensure its adoption.

As a corollary to this proposal, the Department of Justice should streamline its internal MLAT process by having a standard template for MLAT requests so that foreign governments know exactly what information they need to provide to expedite the process. These processes should be further automated and simplified. This is more likely to happen if the White House and State Department take a greater inter-est in reforming the process. MLAT requests could become a larger part of managing bilateral relationships, increasing the visibility and importance of MLAT requests and thus reducing the response time. MLAT requests might also be used as a source of leverage, with access to a streamlined process used as a diplomatic bargaining chip.

# Deterrence

The private sector is the main target of cyberattacks. The government has so far been unable to defend it from industrial espionage or more disruptive attacks, and several federal laws prevent companies from active defense measures or "hacking back" to retaliate against or disrupt attackers. Although companies must improve their own defenses, policies taken to deter the most sophisticated state actors would be an important step in reducing the threats and thus restoring some measure of confidence in the technology sector that the government can effectively address the cybersecurity challenge.

As many international relations experts have noted, cyber deterrence is not the same as Cold War nuclear deterrence.[33] Nuclear deterrence relied on clear attribution; massive, mutual destruction of a limited number of potential adversaries; and a willingness to build up and display deadly weapon systems. The challenges of deterrence in cyberspace are significantly more complex. States have no monopoly on cyber weapons and there are numerous malicious actors. States can choose to outsource hacking and intrusion activities to proxy parties, providing deniability and greater obfuscation. Moreover, the private sector is a frequent target for state-backed attackers seeking political and commercial advantages. Cyberattacks have uncertain consequences and may result in unintended outcomes as they spread to other networks. The disclosure of specific offensive cyber capabilities is likely to lead adversaries to develop effective countermeasures. A number of analysts have argued that security models drawn from public health or climate change are more appropriate than Cold War analogies.[34]

Although the ability to assign responsibility for an attack has significantly improved, high-confidence attribution remains a relatively slow and secretive process. The White House's willingness to name North Korea in December 2014 as the culprit behind the attack on Sony Pictures Entertainment was notable because it occurred only weeks after the intrusion became public. In contrast, it took the U.S. government almost four months to publicly attribute the attacks on the Democratic National Committee to Russia after they were first revealed in June 2016.[35] Even if the actors behind an attack can be identified, their intentions may remain elusive. Some may enter networks to pursue direct commercial espionage, while others lay the groundwork for future attacks. Moreover, the methods and technologies used to identify intrusions and assign responsibility are similar to those deployed in offensive cyber operations. Attribution may not only reveal intelligence techniques used for espionage but also allow a potential adversary to patch vulnerabilities, which could result in the loss of U.S. capabilities.[36]

In a 2015 statement to Congress, the Obama administration identified four types of cyberattacks it would deter: attacks intended to cause casualties; attacks intended to cause significant disruption to the normal functioning of U.S. society or government, including attacks against critical infrastructure; activity that threatens the command and control and the freedom of maneuver of U.S. military forces; and malicious cyber activity that undermines national economic security through economic espionage or sabotage.[37] The United States has so far only been able to deter state-backed destructive attacks on critical infrastructure and attacks that cause deaths or destruction. This deterrence is possible in part because of U.S. offensive capabilities and in part because there appears to be a high degree of self-restraint involved. Threats such as the Islamic State, which might seek mass destabilization of physical and digital infrastructure, evidently do not yet have the capability to do so. Highly capable states such as China and Russia do not want to undermine a digital infrastructure that is increasingly important to their economies, although Russia seemed willing to drop such restraint in its alleged attack on a Ukrainian power grid.

The Trump administration, as it continues discussions with Beijing and Moscow aimed at preventing cyberattacks from escalating into physical conflict, should emphasize the value of self-restraint. To reduce the chances of misperception or miscalculation, the Department of Defense should continue the trend of greater transparency in the doctrine and rules of engagement in cyberspace, as well as the development of confidence-building measures such as crisis hotlines and joint exercises. While increasing the potency and transparency of its offensive capabilities, the United States should signal that it intends to maintain the current policy that only the president has the ability to authorize an offensive cyber operation that would disrupt or destroy.

The vast majority of cyber operations have fallen below the threshold that would trigger an armed conflict, and they have thus not been deterred. As these types of attacks continue and threaten economic, political, and diplomatic interests, the importance of publicly identifying and calling out attackers will increase. The United States should be willing to disclose information about cyberattacks, if only to put states and their proxies on notice and direct the world's attention to online threats, much as it did when it denounced North Korea, China, Russia (see the text box), and Iran.[38]

Once those responsible have been named, the United States will need to target assets valuable to both the attackers and the policymakers who order the actions or, through their inaction, acquiesce to them. Covert cyber operations, especially those designed to dismantle an attacker's offensive infrastructure and disrupt future attacks, should be carefully considered, but such attacks have no deterrent effect and, in fact, risk escalation. Diplomatic actions such as canceling high-level talks, economic sanctions on companies and state entities, travel restrictions, and other visible punitive sanctions on senior officials are more likely to deter state actors. Retaliatory responses will require closer coordination with allies, both on the levying of sanctions and on developing military exercises to ensure the resiliency of defense, communication, transportation, and power networks. Given the tit-for-tat character of cyber conflict, retaliation and sanctions should continually be considered in their strategic context.

## CONTENDING WITH RUSSIA

Russian attempts to undermine the U.S. political process through cyber methods is a qualitative change in behavior. Until the summer of 2016, U.S. officials publicly described Russian hackers as a significant threat, but compared with their Chinese counterparts, Russian hackers were more stealthy and sophisticated, and less likely to leave evidence they had penetrated networks. While criminal networks, nationalistic hackers, and other non-state groups in Russia targeted the private sector, state-sponsored hackers focused on traditional espionage directed at political and military networks as well as critical infrastructure, perhaps to prepare for destructive attacks.

In a joint statement in October 2016, the Director of National Intelligence and the Department of Homeland Security declared that the intelligence community was "confident that the Russian Government directed the recent compromises of e-mails from US persons and institutions, including from US political organizations."[39] The *Washington Post* reported in December 2016 that the CIA had assessed that Russia interfered in the election to tilt the election to Trump, not just undermine confidence in the electoral system.[40] President Obama ordered a "full review" of "hacking-related activity aimed at disrupting" elections that date back to 2008.

In December 2016, the White House announced that it was expelling thirty-five intelligence operatives from the United States and sanctioning nine entities and individuals: two Russian intelligence services, four individual intelligence officers, and three companies that provided material support to cyber operations. The Department of Homeland Security and the FBI released a joint analytic report that provided more details on the hacking. President Obama also suggested there would be covert retaliatory measures directed at Moscow.[41]

The challenge for the White House was to design a response that penalized Russia, but did not risk escalation, the peril of which is much higher for the United States given its greater dependence on the internet. The retaliation also needed to deter future attacks on the United States and its allies but at the same time not under-mine efforts to develop rules of behavior for states in cyberspace. The ejections and sanctions walked this line, but they came too long after the original attack, allowing others to view it as a success and weakening the deterrent effect for future attacks. There is more the United States needs to do, including increasing aid to Estonia, Latvia, Ukraine, and other states on Russia's periphery; fostering closer cooperation with the United Kingdom, Germany, and France on protecting electoral systems from cyberattacks and countering information operations; and increasing funds for the development of next-generation anonymizing tools for organizations that monitor the Kremlin. The United States should also dismantle the infrastructure Russian hackers used to compromise U.S. political institutions to disrupt future cyber operations. This may involve covert activity or more visible steps, such as working with the international network of computer emergency response teams—much as the United States did to counteract the 2011–2013 Iranian distributed denial of service attacks against U.S. banks. In addition, the United States will want to continue working with allies and partners to strengthen the norm of state responsibility, pressuring governments to make sure nonstate actors do not operate out of their territory.

# Encryption

The debate over the uses and proliferation of encryption is the most visible manifestation of the gap between Washington and Silicon Valley. Over the last two years, Apple, WhatsApp, and other companies have rolled out end-to-end encryption on smartphone operating systems, messaging services, and other online communications products.[42] In these products, the data is scrambled through mathematical formulas, and only the owner has the ability to decrypt it. The device manufacturer or service provider cannot access the data, even when presented with a lawful warrant.

With the spread of these products and services, law enforcement and, to a lesser extent, intelligence agencies have warned that individuals under investigation or surveillance are going dark—with agencies unable to wiretap or access data from individuals, organizations, and criminal enterprises, even with a court order.[43] These agencies argue that encryption will prevent the monitoring of communications and hinder the prevention of crime and terrorist attacks as well as the ability to mount investigations after the fact. FBI Director James Comey has noted, for example, that five hundred of the four thousand devices related to investigations the agency undertook between October 2015 and March 2016 "could not be opened by any means."[44]

Faced with an inability to access encrypted data, some federal and local law agencies have called on U.S. technology companies to provide technological ways to bypass encryption, known as exceptional access.[45] In November 2016, Manhattan District Attorney Cyrus Vance Jr. renewed his call for laws that make it possible for the police to access data on locked iPhones. Vance noted that 423 Apple iPhones and iPads had been lawfully seized in Manhattan since October 2014 but remained inaccessible due to default device encryption.[46] Although the White House has said it supports strong encryption, Comey has criticized the decision to provide end-to-end encryption as a "business model" because U.S. technology companies increasingly depend on foreign markets.[47] In 2015, the major U.S. tech companies reported that 59 percent of their revenues came from foreign markets; Intel receives more than 80 percent of its revenue from abroad, Apple close to 60 percent.[48] Burned by the Snowden disclosures, these companies have an incentive to provide high levels of security to global users and assert their independence from the U.S. government.

Civil libertarians, human rights defenders, and technology companies laud encryption as essential to promoting cybersecurity, protecting free speech, and enabling dissidents to operate more safely under repressive regimes.[49] Tech companies have consistently argued that it is not possible to provide exceptional access to data without compromising the security of all users.[50] If and when such tools are introduced in a product, hackers and states will soon find ways of exploiting back doors.[51]

Proponents of strong encryption also argue that malicious actors will easily circumvent U.S. policy. Neither the government nor the private sector in the United States has a monopoly on encryption tools and methods; even if U.S. companies built in back doors, criminals and terrorists could easily use products developed elsewhere. Two-thirds of the nearly nine hundred hardware and software products that incorporate encryption have been built outside the United States.[52]

External events largely drive the debate, as the pendulum swings widely between civil liberty and national security concerns. In the wake of the 2016 terrorist attacks in Germany and France, the two countries asked the European Union to compel internet companies to help decrypt messages as part of terrorism investigations.[53] Moreover, debates and policies in the United States are being watched and referenced in other countries. Beijing and Moscow have their own reasons to demand access to decrypted data, but they are also more than happy to use what Washington does as rhetorical cover. Chinese officials, for example, referenced the debate in the United States and United Kingdom as they justified a new antiterrorism law requiring that technology firms decrypt information when requested by the government.[54]

The Obama administration, after deciding not to pursue a legislative option, relied on persuasion and public pressure.[55] Companies, however, want clarity on policy. The Trump administration will be confronted with the same deadlock. Pursuing a solution that involves requiring U.S. companies to maintain an ability to decrypt data when requested to do so by a court will only exacerbate the tension between

Washington and Silicon Valley, lead to lengthy legal battles with civil rights organizations and U.S. tech companies, and perpetuate a perception in foreign markets that U.S. tech companies are incapable of keeping user and customer data private.

Law enforcement and counterterrorism operations face real challenges and need new capabilities. There are alternative options to back doors that the Trump administration should consider. One option would be to give law enforcement the power to gather data by exploiting existing security flaws in software. Known as lawful hacking, this would give the Justice Department and other agencies the ability to hack into a suspect's smartphone or computer with a court order, such as a warrant.[56] The U.S. government and the FBI already have this power, but the FBI needs to expand its capabilities. A broad policy and legal debate to define the parameters of the hacking, followed by strict judicial oversight, would ensure that lawful hacking is used responsibly, much like the restrictions that already apply to wiretapping. A lawful hacking approach would also require the government to clearly articulate a vulnerabilities equity process that would regulate when the government reveals vulnerabilities in software to vendors.[57] In addition, lawful hacking would also need to be complemented by using metadata and exploiting the data provided by the internet of things to diminish the chance of law enforcement going dark when agencies can no longer access data, even when armed with a court order.[58]

The United States Secret Service, Homeland Security Investigations, Air Force Office of Special Investigations, and other federal law enforcement agencies have significant roles and authority in investigating cybercrime. These agencies, along with the FBI and local law enforcement, need to increase technologically literate staff and devote significantly more resources to encryption and anonymization. The FBI currently has only thirty-nine staff members who deal with encryption and anonymization technologies (eleven of whom are agents), and only $31 million in funding for those activities.[59] Change can be achieved, as it was in the 1990s when organized crime started using disposable phones that hampered wiretaps and law enforcement adapted.[60]

# Recommendations

Despite sustained efforts by the Obama administration on domestic and international policy, the United States has neither deterred other states from launching cyberattacks designed to disrupt, coerce, and influence, nor denied them from achieving their goals by adequately hardening public and private sector network defenses.

Continuing or strengthening existing approaches will not likely match the growing sophistication and scope of the threat. Yet the U.S. government will be unable to create new models of cybersecurity until it closes the political gap with the private sector that emerged during the Obama administration over a number of issues in cyberspace policy. Repairing the rift will not be easy, but there are areas where the two sides can find common ground. The U.S. government should take the following steps:

- continue and expand the work of the U.S. Digital Service and develop a time-limited fellowship service specifically focused on cybersecurity, modeled on the CDC's Epidemic Intelligence Service
- amend provisions of the Electronic Communications Privacy Act, using the U.S.-UK agreement as a template, to allow technology companies to provide data to foreign governments, provided the requesting government has a legitimate interest in the criminal activity being investigated, the target is located outside the United States, and the target is not a U.S. person
- streamline the Department of Justice's internal MLAT process by having a standard template for MLAT requests and further automating the process
- attribute attacks more frequently and, for cyberattacks that fall below the use of force and armed attack threshold, devise and implement forceful responses, such as covert cyber operations designed to

disrupt future attacks; cancel high-level talks, impose economic sanctions on companies and state entities, travel restrictions, and other visible punitive sanctions on senior intelligence officers or other high-level officials; and bolster the offensive and defensive cyber capabilities of friends and allies

- continue discussions with Beijing and Moscow aimed at both preventing cyberattacks from escalating into physical conflict and developing confidence-building measures such as crisis hotlines and joint exercises; increase Defense Department transparency in doctrine and cyber rules of engagement; and signal civilian control over offensive cyber operations

- strengthen law enforcement's ability to conduct lawful hacking under strict judicial oversight and a clearly articulated vulnerabilities equity process

- increase funding, staffing, and technology resources at law enforcement agencies dedicated to encryption and anonymization

There is a chance that the relationship between Silicon Valley and Washington could significantly worsen over the next few years, driven by economic interests, policy differences, and President Trump's apparent willingness to call out and criticize the business practices of individual companies. If that happens, the question will not be how best to construct a new approach to cybersecurity. Rather, by default a new approach will emerge because the old consensus on the need for private sector leadership and the importance of developing public-private partnerships will no longer be sustainable.

What comes next is uncertain. The U.S. government may take a much more activist role through regulation, the elevation of the Department of Defense as the lead organization in protecting critical infrastructure, and increased surveillance of domestic networks. The private sector may in turn respond with limited cooperation on information sharing, a greater focus on encryption and other technological solutions to defending their own networks, and individual efforts to make deals with governments around the world to smooth access to technology.

Tension is inevitable, but pragmatism from both sides can contain it. Cybersecurity talent development and forging a framework to fight data localization are two areas where interests converge relatively easily. Constructing a viable deterrence is more difficult, and balancing the needs of lawful access with the increasing use of encryption is even harder, but such an approach would do the most to reduce the discord in the relationship. Hopefully both sides will recognize that an exacerbation of the division between the public and private sector is bound to worsen cybersecurity and diminish the U.S. ability to shape cyberspace.

# Endnotes

1. Tal Kopan, "Cybercrime Costs $575B Yearly," *Politico*, June 10, 2014, http://www
   .politico.com/story/2014/06/cybercrime-yearly-costs-107601.
2. Greg Otto, "Ransomware Attacks Quadrupled in Q1 2016," *Fedscoop*, April 29, 2016,
   http://fedscoop.com/ransomware-attacks-up-300-percent-in-first-quarter-of-2016.
3. Joe Davidson, "Cyberattacks on Personal Health Records Growing 'Exponentially,'"
   *Washington Post*, September 28, 2016, http://www.washingtonpost.com/news/powerpost
   /wp/2016/09/28/cyberattacks-on-personal-health-records-growing-exponentially.
4. Kim Zetter, "Inside the Cunning, Unprecedented Hack of Ukraine's Power Grid," *Wired*,
   March 3, 2016, http://www.wired.com/2016/03/inside-cunning-unprecedented-hack
   -ukraines-power-grid/; Ellen Nakashima, "Russian Government Hackers Penetrated
   DNC, Stole Opposition Research on Trump," *Washington Post*, June 14, 2016, http://
   www.washingtonpost.com/world/national-security/russian-government-hackers
   -penetrated-dnc-stole-opposition-research-on-trump/2016/06/14/cf006cb4-316e
   -11e6-8ff7-7b6c1998b7a0_story.html; Evan Perez, Shimon Prokupecz, and Wesley
   Bruer, "Feds Believe Russians Hacked Florida Election-Systems Vendor," CNN, October 12, 2016, http://www.cnn.com/2016/10/12/politics/florida-election-hack/; Joon
   Ian Wang, "A 'Nation-State' Used Wikileaks to Influence the US Election, the Head
   of the NSA Says," *Quartz*, November 15, 2016, http://qz.com/838615/nsa-chief-on
   -wikileaks-and-the-hacks-affecting-the-us-election-a-conscious-effort-by-a-nation
   -state.
5. David Perera and Joseph Marks, "Newly Disclosed Hacks Got 'Crown Jewels,'" *Polit-
   ico*, June 12, 2015, http://www.politico.com/story/2015/06/hackers-federal-employees
   -security-background-checks-118954.html.
6. Thom Shanker and David Sanger, "U.S. Suspects Iran Was Behind a Wave of Cyberattacks," *New York Times*, October 13, 2012, http://www.nytimes.com/2012/10/14
   /world/middleeast/us-suspects-iranians-were-behind-a-wave-of-cyberattacks
   .html; Devlin Barrett and Danny Yadron, "North Korean Role in Sony Hack Presents Quandary for U.S.," *Wall Street Journal*, December 17, 2014, http://www.wsj.com
   /articles/u-s-has-concluded-north-korea-is-behind-sony-hack-1418861023.
7. Gartner, "Gartner Says 6.4 Billion Connected 'Things' Will Be in Use in 2016, Up
   30 Percent From 2015," November 10, 2015, http://www.gartner.com/newsroom
   /id/3165317.
8. James R. Clapper, "Statement for the Record on the Worldwide Threat Assessment
   of the U.S. Intelligence Community, Senate Armed Services Committee," February
   9, 2016, http://www.dni.gov/files/documents/SASC_Unclassified_2016_ATA_SFR
   _FINAL.pdf.
9. James B. Comey and Sally Quillian Yates, "Going Dark: Encryption, Technology, and
   the Balances Between Public Safety and Privacy," Joint statement before the Senate
   Judiciary Committee, July 8, 2015, http://www.fbi.gov/news/testimony/going-dark

-encryption-technology-and-the-balances-between-public-safety-and-privacy; *Don't Panic: Making Progress on the 'Going Dark' Debate*, Berkman Center for Internet & Society, Harvard University, February 1, 2016, http://cyber.law.harvard.edu/pubrelease /dont-panic/Dont_Panic_Making_Progress_on_Going_Dark_Debate.pdf.

10. Cat Zakrzewski, "Tech Companies Line Up Behind Surveillance Reform Bill," *Tech Crunch*, April 29, 2015, http://techcrunch.com/2015/04/29/tech-companies-line-up -behind-surveillance-reform-bill/; Adam Segal, "Do Local Laws Belong in a Global Cloud? Q&A With Brad Smith of Microsoft (Part One)," *Net Politics*, August 26, 2015, http://blogs.cfr.org/cyber/2015/08/26/do-local-laws-belong-in-a-global-cloud-qa -with-brad-smith-of-microsoft-part-one/; Amul Kalia, "Where Do Major Tech Companies Stand on Encryption?" *Electronic Frontier Foundation*, October 9, 2015, http:// www.eff.org/deeplinks/2015/10/where-do-major-tech-companies-stand-encryption.

11. David Kravets, "Trump Urges Supporters to Boycott Apple in Wake of Encryption Brouhaha," *Ars Technica*, February 19, 2016, http://arstechnica.com/tech-policy/2016 /02/trump-urges-supporters-to-boycott-apple-in-wake-of-encryption-brouhaha.

12. "International Strategy for Cyberspace: Prosperity, Security, and Openness in a Networked World," Office of the President of the United States, May 2011, http://www/ .whitehouse.gov/sites/default/files/rss_viewer/internationalstrategy_cyberspace .pdf.; "Fact Sheet: Cybersecurity National Action Plan," Office of the Press Secretary, White House, February 9, 2016, http://www.whitehouse.gov/the-press -office/2016/02/09/fact-sheet-cybersecurity-national-action-plan; "Executive Order— Promoting Private Sector Cybersecurity Information Sharing," Office of the Press Secretary, White House, Washington, DC, February 13, 2015, http://www.whitehouse .gov/the-press-office/2015/02/13/executive-order-promoting-private-sector -cybersecurity-information-shari; "Executive Order—Blocking the Property of Certain Persons Engaging in Significant Malicious Cyber-Enabled Activities," Office of the Press Secretary, White House, Washington, DC, April 1, 2015, http://www.whitehouse .gov/the-press-office/2015/04/01/executive-order-blocking-property-certain -persons-engaging-significant-m; Ellen Nakashima and Missy Ryan, "U.S. Military Has Launched a New Digital War Against the Islamic State," *Washington Post*, July 15, 2016, http://www.washingtonpost.com/world/national-security/us-militarys-digital -war-against-the-islamic-state-is-off-to-a-slow-start/2016/07/15/76a3fe82-3da3 -11e6-a66f-aa6c1883b6b1_story.html.

13. "President Xi Jinping's State Visit to the United States," Fact Sheet, Office of the Press Secretary, White House, September 25, 2015, http://www.whitehouse.gov /the-press-office/2015/09/25/fact-sheet-president-xi-jinpings-state-visit-united-states.

14. FireEye iSight Intelligence, "Red Line Drawn: China Recalculates Its Use of Cyber Espionage," June 2016, http://www.fireeye.com/content/dam/fireeye-www/current -threats/pdfs/rpt-china-espionage.pdf; Joe Uchill, "Obama Administration Confirms Drop in Chinese Cyber Attacks," *The Hill*, June 28, 2016, http://thehill.com/policy /cybersecurity/285153-obama-administration-confirms-drop-in-chinese-cyber -attacks.

15. Andrea Shalal, "Top U.S. Spy Says Skeptical About U.S.-China Cyber Agreement," Reuters, September 30, 2015, http://www.reuters.com/article/us-usa-cybersecurity -idUSKCN0RT1Q820150930; Jack Goldsmith, "China and Cybertheft: Did Action Follow Words," *Lawfare*, March 18, 2016, http://www.lawfareblog.com/china-and -cybertheft-did-action-follow-words.

16. "Report of the Group of Governmental Experts on Developments in the Field of Information and Telecommunications in the Context of International Security," United Nations General Assembly, July 22, 2015, http://www.un.org/ga/search/view_doc .asp?symbol=A/70/174.

17.  Jessica Silver-Greenberg, Matthew Goldstein, and Nicole Perlroth, "JPMorgan Chase Hacking Affects 76 Million Households," *New York Times*, October 2, 2014, http://dealbook.nytimes.com/2014/10/02/jpmorgan-discovers-further-cyber-security -issues/?_r=0; Steve Morgan, "Why J.P. Morgan Chase & Co. Is Spending a Half Billion Dollars on Cybersecurity," *Forbes*, January 30, 2016, http://www.forbes.com /sites/stevemorgan/2016/01/30/why-j-p-morgan-chase-co-is-spending-a-half-billion -dollars-on-cybersecurity/#5044bdcc2a7f.

18.  Kim S. Nash, "2010 State of the CIO: Today's Focus for IT Departments—Business Opportunities," *CIO*, December 17, 2009, http://www.cio.com/article/2421968/strategy /2010-state-of-the-cio--today-s-focus-for-it-departments---business-opportunities.html; "2016 State of the CIO Survey," *CIO*, January 2016, reformhttp://coreo.staticworld .net/assets/2016/01/14/2016-state-of-the-cio-executive-summary.pdf.

19.  Piers Wilson, "Security Market Trends and Predictions," IISP, http://iisp.informz.net /IISP/data/images/WhitePaperWebsite.pdf.

20.  "Hacking the Skills Shortage: A study of the international shortage in cybersecu- rity skills," Center for Strategic and International Studies, 2016, https://csis-prod .s3.amazonaws.com/s3fs-public/160727_rpt_cybersecurity_workforce_shortage.pdf.

21.  Ibid.

22.  "Transforming Government Services Through Technology and Innovation," Impact Report, Office of the Press Secretary, White House, August 9, 2016, https:// www.whitehouse.gov/the-press-office/2016/08/09/impact-report-transforming -government-services-through-technology-and; Patrick Tucker, "Meet the Head of the Pentagon's Agile New Digital Team," *Defense One*, November 18, 2015, http:// www.defenseone.com/technology/2015/11/meet-head-pentagons-agile-new-digital -service/123825/; Jim Garamone, "Defense Digital Service Chief Brings Private- Sector Experience to Job," *DoD News*, June 10, 2016, http://www.defense.gov/News /Article/Article/796409/defense-digital-service-chief-brings-private-sector -expertise-to-job; Lisa Ferdinando, "Carter Announces 'Hack the Pentagon' Program Results," *DoD News*, June 17, 2016, http://www.defense.gov/News/Article/Article /802828/carter-announces-hack-the-pentagon-program-results.

23.  See, for example, Elizabeth Segran, "Can Obama's Tech Legacy Survive Trump?" *Fast Company*, December 7, 2016, http://www.fastcompany.com/3062477 /innovation-agents/can-obamas-tech-legacy-survive-president-trump.

24.  Jack Moore, "Bill Would Extend U.S. Digital Service Beyond the Obama Admin- istration," *Nextgov*, May 31, 2016, http://www.nextgov.com/cio-briefing/2016/05 /bill-would-extend-us-digital-service-beyond-obama-administration/128695.

25.  Loretta Chao and Paulo Trevisani, "Brazil Legislators Bear Down on Inter- net Bill," *Wall Street Journal*, November 13, 2013, http://www.wsj.com/articles /SB10001424052702304868404579194290325348688; Allison Grande, "Brazil Nixes Data Localization Mandate From Internet Bill," *Law 360*, March 20, 2014, http:// www.law360.com/articles/520198/brazil-nixes-data-localization-mandate-from -internet-bill; Joe McDonald, "Business Groups Appeal to China Over Cyber- security Law," Associated Press, August 11, 2015, http://www.washingtonpost.com /world/asia_pacific/business-groups-appeal-to-china-over-cybersecurity-law/2016 /08/11/b99cd610-6039-11e6-84c1-6d27287896b5_story.html; Gillian Wong, "China Cybersecurity Fears Prompt Business Groups to Press Obama," *Wall Street Journal*, August 12, 2015, http://www.wsj.com/articles/business-tech-groups-press-obama-on -china-competition-concerns-1439361855.

26.  Laura Mills, "New Russian Data Laws Worry Rights Activists, Telecom Compa- nies," *Wall Street Journal*, July 7, 2016, http://www.wsj.com/articles/new-russian-data -laws-worry-rights-activists-telecom-companies-1467905452; Adam Taylor, "Russia

Moves to Block Professional Networking Site LinkedIn," *Washington Post*, November 17, 2016, http://www.washingtonpost.com/news/worldviews/wp/2016/11/17/russia-moves-to-block-professional-networking-site-linkedin/.

27. "Electronic Communications Privacy Act of 1986 (ECPA)," Justice Information Sharing, http://it.ojp.gov/privacyliberty/authorities/statutes/1285.

28. Gail Kent, "The Mutual Legal Assistance Problem Explained," Center for Internet and Society, February 23, 2015, http://cyberlaw.stanford.edu/blog/2015/02/mutual-legal-assistance-problem-explained; Peter Swire and Justin D. Hemmings, "Re-Engineering the Mutual Legal Assistance Treaty Process," paper presented at the NYU Law and PLSC Conferences, May 14, 2015.

29. Matthias Bauer, Martina F. Ferracane, and Erik van der Marel, "Tracing the Economic Impact of Regulations on the Free Flow of Data and Data Localization," *Global Commission on Internet Governance*, CIGI, May 2016, http://www.cigionline.org/sites/default/files/gcig_no30web_2.pdf.

30. Jennifer Daskal, "A New UK-US Data Sharing Agreement: A Tremendous Opportunity, if Done Right," *Just Security*, February 8, 2016, https://www.justsecurity.org/29203/british-searches-america-tremendous-opportunity.

31. Peter J. Kadzik, "Letter From the Assistant Attorney General to the President of the United States Senate," July 15, 2016, http://www.documentcloud.org/documents/2994379-2016-7-15-US-UK-Biden-With-Enclosures.html#document/p1.

32. David Kris, "U.S. Government Presents Draft Legislation for Cross Border Data Requests," *Lawfare*, July 16, 2016, http://www.lawfareblog.com/us-government-presents-draft-legislation-cross-border-data-requests; Jennifer Daskal and Andrew K. Woods, "Congress Should Embrace the DOJ's Cross-Border Data Fix," *Just Security,* August 1, 2016, http://www.justsecurity.org/32213/congress-embrace-dojs-cross-border-data-fix.

33. Martin C. Libicki, *Cyberdeterrence and Cyberwar* (Santa Monica, CA: RAND Corporation, 2009); Stephen Blank, "Can Information Warfare Be Deterred?" in David S. Alberts and Daniel S. Papp, *Information Age Anthology*. Volume III, *The Information Age Military* (Washington, DC: Command and Control Research Program, 2001), 125-57; Will Goodman, "Cyber Deterrence: Tougher in Theory Than in Practice?" *Strategic Studies Quarterly* (Fall 2010), http://www.au.af.mil/au/ssq/2010/fall/goodman.pdf; David Elliott, "Deterring Strategic Cyberattack," *IEEE Security & Privacy* (September/October 2011), 36–39.

34. Gregory J. Rattray and Jason Healey, "Non-State Actors and Cyber Conflict, in America's Cyber Future: Security and Prosperity in the Information Age," Center for a New American Security; see also, "Barack Obama, Neural Nets, Self Driving Cars, and the Future of the World," *Wired*, http://www.wired.com/2016/10/president-obama-mit-joi-ito-interview.

35. Nakashima, "Russian Government Hackers Penetrated DNC, Stole Opposition Research on Trump;" "Joint Statement From the Department of Homeland Security and Office of the Director of National Intelligence on Election Security," October 7, 2016, http://www.dni.gov/index.php/newsroom/press-releases/215-press-releases-2016/1423-joint-dhs-odni-election-security-statement.

36. Sean Kanuck, "Statement for the Record for the House Committee on Oversight and Government Reform," Subcommittees on Information Technology and National Security, July 13, 2016, http://oversight.house.gov/wp-content/uploads/2016/07/Kanuck-Statement-Digital-Acts-of-War-7-13.pdf.

37. "Report to Congress on Cyber Deterrence," Inside Cyber Security, http://insidecybersecurity.com/sites/insidecybersecurity.com/files/documents/dec2015/cs2015_0133.pdf.

38. Carol E. Lee and Jay Solomon, "U.S. Targets North Korea in Retaliation for Sony Hack," *Wall Street Journal*, January 3, 2015, http://www.wsj.com/articles/u-s-penalizes-north-korea-in-retaliation-for-sony-hack-1420225942; Michael S. Schmidt and David E. Sanger, "5 in China Army Face U.S. Charges of Cyberattacks," *New York Times*, May 19, 2014, http://www.nytimes.com/2014/05/20/us/us-to-charge-chinese-workers-with-cyberspying.html?_r=0; Ellen Nakashima and Matt Zapotosky, "U.S. Charges Iran-Linked Hackers With Targeting Banks, N.Y. Dam," *Washington Post*, March 24, 2016, http://www.washingtonpost.com/world/national-security/justice-department-to-unseal-indictment-against-hackers-linked-to-iranian-government/2016/03/24/9b3797d2-f17b-11e5-a61f-e9c95c06edca_story.html.

39. Director of National Intelligence, Joint Statement From the Department of Homeland Security and Office of the Director of National Intelligence on Election Security, October 7, 2016, Washington, DC, http://www.dni.gov/index.php/newsroom/press-releases/215-press-releases-2016/1423-joint-dhs-odni-election-security-statement.

40. Adam Entous, Ellen Nakashima, and Greg Miller, "Secret CIA Assessment Says Russia Was Trying to Help Trump Win White House," *Washington Post*, December 9, 2016, http://www.washingtonpost.com/world/national-security/obama-orders-review-of-russian-hacking-during-presidential-campaign/2016/12/09/31d6b300-be2a-11e6-94ac-3d324840106c_story.html?utm_term=.eb85ff12c75c

41. Statement by the President on Actions in Response to Russian Malicious Cyber Activity and Harassment, December 29, 2016, http://www.whitehouse.gov/the-press-office/2016/12/29/statement-president-actions-response-russian-malicious-cyber-activity.

42. John Zorabedian, "Apple iMessage's End-to-End Encryption Stymies US Data Request," *Naked Security*, September 10, 2015, http://nakedsecurity.sophos.com/2015/09/10/apple-imessages-end-to-end-encryption-stymies-us-data-request/; Cade Metz, "Forget Apple vs. the FBI: WhatsApp Just Switched On Encryption for a Billion People," *Wired*, April 5, 2016, http://www.wired.com/2016/04/forget-apple-vs-fbi-whatsapp-just-switched-encryption-billion-people/.

43. James B. Comey and Sally Quillian Yates, "Going Dark: Encryption, Technology, and the Balances Between Public Safety and Privacy," joint statement before the Senate Judiciary Committee, July 8, 2015, https://www.fbi.gov/news/testimony/going-dark-encryption-technology-and-the-balances-between-public-safety-and-privacy.

44. Kim S. Nash, "FBI Director Says Agency Is Talking With Tech Firms About Privacy, Encryption," *Wall Street Journal*, July 27, 2016, http://blogs.wsj.com/cio/2016/07/27/fbi-director-says-agency-is-talking-with-tech-firms-about-privacy-encryption/.

45. Cyrus R. Vance Jr., François Molins, Adrian Leppard, and Javier Zaragoza, "When Phone Encryption Blocks Justice," *New York Times*, August 11, 2015, http://www.nytimes.com/2015/08/12/opinion/apple-google-when-phone-encryption-blocks-justice.html.

46. "Report of the Manhattan District Attorney's Office on Smartphone Encryption and Public Safety," November 2016, http://manhattanda.org/sites/default/files/Report%20on%20Smartphone%20Encryption%20and%20Public%20Safety:%20An%20Update.pdf

47. Dan Froomkin and Jenna McLaughlin, "Comey Calls on Tech Companies Offering End-to-End Encryption to Reconsider 'Their Business Model,'" *The Intercept*, December 9, 2015, https://theintercept.com/2015/12/09/comey-calls-on-tech-companies-offering-end-to-end-encryption-to-reconsider-their-business-model.

48. Matt Kranz, "10 U.S. Companies Take the Most Foreign Money," *USA Today*, July 15, 2015, http://americasmarkets.usatoday.com/2015/07/15/10-u-s-companies-take-the-most-foreign-money.

49. Electronic Frontier Foundation, "Anonymity and Encryption: Comments Submitted to the United Nations Special Rapporteur on the Promotion and Protection of the Right to Freedom of Opinion and Expression," February 10, 2015, http://www.eff.org/files/2015/02/10/unanonymity-encryption-eff.pdf.

50. Access et al., "Open Letter on Encryption to President Barack Obama," May 19, 2015, https://static.newamerica.org/attachments/3138--113/Encryption_Letter_to_Obama_final_051915.pdf.

51. Harold Abelson et al., "Keys Under Doormats: Mandating Insecurity by Requiring Government Access to All Data and Communications," MIT Computer Science and Artificial Intelligence Laboratory Technical Report, July 6, 2015, http://dspace.mit.edu/handle/1721.1/97690.

52. Bruce Schneier, Kathleen Seidel, and Saranya Vijayakumar, "A Worldwide Survey of Encryption Products," February 11, 2016, http://www.schneier.com/academic/paperfiles/worldwide-survey-of-encryption-products.pdf.

53. Noemie Bisserbe and Natalia Drozdiak, "France, Germany Push for Access to Private Internet Messages in Terror Probes," *Wall Street Journal*, August 23, 2016, http://www.wsj.com/articles/france-germany-push-for-access-to-private-internet-messages-in-terror-probes-1471976681.

54. Ben Blanchard, "China Passes Controversial Counter-Terrorism Law," Reuters, December 28, 2015, http://www.reuters.com/article/us-china-security-idUSKBN0UA07220151228.

55. Cat Zakrzewski, "Encrypted Smartphones Challenge Investigators," *Wall Street Journal,* October 12, 2015, http://www.wsj.com/articles/encrypted-smartphones-challenge-investigators-1444692995.

56. Steven M. Bellovin et al., "Lawful Hacking: Using Existing Vulnerabilities for Wiretapping on the Internet," *Northwestern Journal of Technology and Intellectual Property* 12, no. 1 (2014), http://scholarlycommons.law.northwestern.edu/cgi/viewcontent.cgi?article=1209&context=njtip.

57. Susan Hennessey, "Lawful Hacking and the Case for a Strategic Approach to 'Going Dark,'" Brookings, October 7, 2016, http://www.brookings.edu/research/lawful-hacking-and-the-case-for-a-strategic-approach-to-going-dark.

58. *Don't Panic: Making Progress on the "Going Dark" Debate.*

59. Robyn Greene, "Unbounded and Unpredictable," Open Technology Institute, New America, August 22, 2016, https://www.newamerica.org/oti/blog/unbounded-and-unpredictable.

60. David Simon, "We Are Shocked, Shocked...," *The Audacity of Despair*, June 7, 2013, http://davidsimon.com/we-are-shocked-shocked.

# About the Author

**Adam Segal** is the Ira A. Lipman chair in emerging technologies and national security and director of the Digital and Cyberspace Policy program at the Council on Foreign Relations. An expert on security issues, technology development, and Chinese domestic and foreign policy, Segal was the project director for the CFR-sponsored Independent Task Force report *Defending an Open, Global, Secure, and Resilient Internet*. His most recent book, *The Hacked World Order: How Nations Fight, Trade, Maneuver, and Manipulate in the Digital Age,* describes the increasingly contentious geopolitics of cyberspace. His work has appeared in the *Financial Times*, the *Economist*, *Foreign Policy*, the *Wall Street Journal*, and *Foreign Affairs*, among others. He currently writes for the *Net Politics* blog.

Before coming to CFR, Segal was an arms control analyst for the China Project at the Union of Concerned Scientists. There, he wrote about missile defense, nuclear weapons, and Asian security issues. He has been a visiting scholar at the Hoover Institution at Stanford University, the Massachusetts Institute of Technology's Center for International Studies, the Shanghai Academy of Social Sciences, and Tsinghua University in Beijing. He has taught at Vassar College and Columbia University. Segal is the author of *Advantage: How American Innovation Can Overcome the Asian Challenge* and *Digital Dragon: High-Technology Enterprises in China*, as well as several articles and book chapters on Chinese technology policy.

Segal has a BA and PhD in government from Cornell University, and an MA in international relations from the Tufts University's Fletcher School of Law and Diplomacy.

# Advisory Committee for
## *Rebuilding Trust Between Silicon Valley and Washington*

Dmitri Alperovitch
*CrowdStrike*

Robert O. Boorstin
*Albright Stonebridge Group*

Jeff A. Brueggeman
*AT&T*

Larry Brilliant
*Skoll Global Threats Fund*

Michael Chertoff
*The Chertoff Group*

Steven A. Denning
*General Atlantic, LLC*

Larry J. Diamond
*Hoover Institution*

Eileen Donahoe
*Human Rights Watch*

William H. Draper
*Draper International*

David P. Fidler
*Council on Foreign Relations*

Martha Finnemore
*George Washington University*

Celeste V. Ford
*Stellar Solutions, Inc.*

Tressa S. Guenov
*Department of Defense*

Reid Hoffman
*LinkedIn*

Eugene J. Huang
*Brookings Institution*

William E. Hudson
*Google*

Sean Joyce
*PricewaterhouseCoopers, LLP*

Robert K. Knake
*Council on Foreign Relations*

Anthony P. Lee
*Altos Ventures*

Herbert Lin
*Center for International Security and Cooperation*

Catherine B. Lotrionte
*Georgetown University*

Naotaka Matsukata
*FairWinds Partners, LLC*

Annie Maxwell
*Skoll Global Threats Fund*

Angela McKay
*Microsoft Corporation*

Jeff Moss
*DEF CON Communications*

Craig James Mundie*
*Mundie & Associates*

Matthew Perault
*Facebook*

Robert Pepper
*Facebook*

Neal A. Pollard
*PricewaterhouseCoopers, LLP*

Don Proctor
*Cisco Systems, Ltd.*

Philip R. Reitinger
*VisionSpear, LLC*

Thomas J. Ridge
*Ridge Global, LLC*

Peter Schwartz
*Salesforce.com*

Matthew Spence
*Andreessen Horowitz*

Raj Shah
*Defense Innovation Unit Experimental*

Philip John Venables
*The Goldman Sachs Group*

Amy E. Weaver
*Salesforce.com*

*Co-chair of the advisory committee

# Council Special Reports

*Published by the Council on Foreign Relations*

*Ending South Sudan's Civil War*
Kate Almquist Knopf; CSR No. 77, November 2016
A Center for Preventive Action Report

*Repairing the U.S.-Israel Relationship*
Robert D. Blackwill and Philip H. Gordon; CSR No. 76, November 2016

*Securing a Democratic Future for Myanmar*
Priscilla A. Clapp; CSR No. 75, March 2016
A Center for Preventive Action Report

*Xi Jinping on the Global Stage: Chinese Foreign Policy Under a Powerful but Exposed Leader*
Robert D. Blackwill and Kurt M. Campbell; CSR No. 74, February 2016
An International Institutions and Global Governance Program Report

*Enhancing U.S. Support for Peace Operations in Africa*
Paul D. Williams; CSR No. 73, May 2015

*Revising U.S. Grand Strategy Toward China*
Robert D. Blackwill and Ashley J. Tellis; CSR No. 72, March 2015
An International Institutions and Global Governance Program Report

*Strategic Stability in the Second Nuclear Age*
Gregory D. Koblentz; CSR No. 71, November 2014

*U.S. Policy to Counter Nigeria's Boko Haram*
John Campbell; CSR No. 70, November 2014
A Center for Preventive Action Report

*Limiting Armed Drone Proliferation*
Micah Zenko and Sarah Kreps; CSR No. 69, June 2014
A Center for Preventive Action Report

*Reorienting U.S. Pakistan Strategy: From Af-Pak to Asia*
Daniel S. Markey; CSR No. 68, January 2014

*Afghanistan After the Drawdown*
Seth G. Jones and Keith Crane; CSR No. 67, November 2013
A Center for Preventive Action Report

Council Special Reports 39

*Somalia: A New Approach*
Bronwyn E. Bruton; CSR No. 52, March 2010
A Center for Preventive Action Report

*The Future of NATO*
James M. Goldgeier; CSR No. 51, February 2010
An International Institutions and Global Governance Program Report

*The United States in the New Asia*
Evan A. Feigenbaum and Robert A. Manning; CSR No. 50, November 2009
An International Institutions and Global Governance Program Report

*Intervention to Stop Genocide and Mass Atrocities: International Norms and U.S. Policy*
Matthew C. Waxman; CSR No. 49, October 2009
An International Institutions and Global Governance Program Report

*Enhancing U.S. Preventive Action*
Paul B. Stares and Micah Zenko; CSR No. 48, October 2009
A Center for Preventive Action Report

*The Canadian Oil Sands: Energy Security vs. Climate Change*
Michael A. Levi; CSR No. 47, May 2009
A Maurice R. Greenberg Center for Geoeconomic Studies Report

*The National Interest and the Law of the Sea*
Scott G. Borgerson; CSR No. 46, May 2009

*Lessons of the Financial Crisis*
Benn Steil; CSR No. 45, March 2009
A Maurice R. Greenberg Center for Geoeconomic Studies Report

*Global Imbalances and the Financial Crisis*
Steven Dunaway; CSR No. 44, March 2009
A Maurice R. Greenberg Center for Geoeconomic Studies Report

*Eurasian Energy Security*
Jeffrey Mankoff; CSR No. 43, February 2009

*Preparing for Sudden Change in North Korea*
Paul B. Stares and Joel S. Wit; CSR No. 42, January 2009
A Center for Preventive Action Report

*Averting Crisis in Ukraine*
Steven Pifer; CSR No. 41, January 2009
A Center for Preventive Action Report

*Congo: Securing Peace, Sustaining Progress*
Anthony W. Gambino; CSR No. 40, October 2008
A Center for Preventive Action Report

*Deterring State Sponsorship of Nuclear Terrorism*
Michael A. Levi; CSR No. 39, September 2008

*China, Space Weapons, and U.S. Security*
Bruce W. MacDonald; CSR No. 38, September 2008

*Sovereign Wealth and Sovereign Power: The Strategic Consequences of American Indebtedness*
Brad W. Setser; CSR No. 37, September 2008
A Maurice R. Greenberg Center for Geoeconomic Studies Report

*Securing Pakistan's Tribal Belt*
Daniel S. Markey; CSR No. 36, July 2008 (web-only release) and August 2008
A Center for Preventive Action Report

*Avoiding Transfers to Torture*
Ashley S. Deeks; CSR No. 35, June 2008

*Global FDI Policy: Correcting a Protectionist Drift*
David M. Marchick and Matthew J. Slaughter; CSR No. 34, June 2008
A Maurice R. Greenberg Center for Geoeconomic Studies Report

*Dealing with Damascus: Seeking a Greater Return on U.S.-Syria Relations*
Mona Yacoubian and Scott Lasensky; CSR No. 33, June 2008
A Center for Preventive Action Report

*Climate Change and National Security: An Agenda for Action*
Joshua W. Busby; CSR No. 32, November 2007
A Maurice R. Greenberg Center for Geoeconomic Studies Report

*Planning for Post-Mugabe Zimbabwe*
Michelle D. Gavin; CSR No. 31, October 2007
A Center for Preventive Action Report

*The Case for Wage Insurance*
Robert J. LaLonde; CSR No. 30, September 2007
A Maurice R. Greenberg Center for Geoeconomic Studies Report

*Reform of the International Monetary Fund*
Peter B. Kenen; CSR No. 29, May 2007
A Maurice R. Greenberg Center for Geoeconomic Studies Report

*Nuclear Energy: Balancing Benefits and Risks*
Charles D. Ferguson; CSR No. 28, April 2007

*Nigeria: Elections and Continuing Challenges*
Robert I. Rotberg; CSR No. 27, April 2007
A Center for Preventive Action Report

*The Economic Logic of Illegal Immigration*
Gordon H. Hanson; CSR No. 26, April 2007
A Maurice R. Greenberg Center for Geoeconomic Studies Report

*The United States and the WTO Dispute Settlement System*
Robert Z. Lawrence; CSR No. 25, March 2007
A Maurice R. Greenberg Center for Geoeconomic Studies Report

*Bolivia on the Brink*
Eduardo A. Gamarra; CSR No. 24, February 2007
A Center for Preventive Action Report

*After the Surge: The Case for U.S. Military Disengagement From Iraq*
Steven N. Simon; CSR No. 23, February 2007

*Darfur and Beyond: What Is Needed to Prevent Mass Atrocities*
Lee Feinstein; CSR No. 22, January 2007

*Avoiding Conflict in the Horn of Africa: U.S. Policy Toward Ethiopia and Eritrea*
Terrence Lyons; CSR No. 21, December 2006
A Center for Preventive Action Report

*Living with Hugo: U.S. Policy Toward Hugo Chávez's Venezuela*
Richard Lapper; CSR No. 20, November 2006
A Center for Preventive Action Report

*Reforming U.S. Patent Policy: Getting the Incentives Right*
Keith E. Maskus; CSR No. 19, November 2006
A Maurice R. Greenberg Center for Geoeconomic Studies Report

*Foreign Investment and National Security: Getting the Balance Right*
Alan P. Larson and David M. Marchick; CSR No. 18, July 2006
A Maurice R. Greenberg Center for Geoeconomic Studies Report

*Challenges for a Postelection Mexico: Issues for U.S. Policy*
Pamela K. Starr; CSR No. 17, June 2006 (web-only release) and November 2006

*U.S.-India Nuclear Cooperation: A Strategy for Moving Forward*
Michael A. Levi and Charles D. Ferguson; CSR No. 16, June 2006

*Generating Momentum for a New Era in U.S.-Turkey Relations*
Steven A. Cook and Elizabeth Sherwood-Randall; CSR No. 15, June 2006

*Peace in Papua: Widening a Window of Opportunity*
Blair A. King, CSR No. 14, March 2006
A Center for Preventive Action Report

*Neglected Defense: Mobilizing the Private Sector to Support Homeland Security*
Stephen E. Flynn and Daniel B. Prieto; CSR No. 13, March 2006

*Afghanistan's Uncertain Transition From Turmoil to Normalcy*
Barnett R. Rubin; CSR No. 12, March 2006
A Center for Preventive Action Report

*Preventing Catastrophic Nuclear Terrorism*
Charles D. Ferguson; CSR No. 11, March 2006

*Getting Serious About the Twin Deficits*
Menzie D. Chinn; CSR No. 10, September 2005
A Maurice R. Greenberg Center for Geoeconomic Studies Report

*Both Sides of the Aisle: A Call for Bipartisan Foreign Policy*
Nancy E. Roman; CSR No. 9, September 2005

*Forgotten Intervention? What the United States Needs to Do in the Western Balkans*
Amelia Branczik and William L. Nash; CSR No. 8, June 2005
A Center for Preventive Action Report

*A New Beginning: Strategies for a More Fruitful Dialogue with the Muslim World*
Craig Charney and Nicole Yakatan; CSR No. 7, May 2005

*Power-Sharing in Iraq*
David L. Phillips; CSR No. 6, April 2005
A Center for Preventive Action Report

*Giving Meaning to "Never Again": Seeking an Effective Response to the Crisis
in Darfur and Beyond*
Cheryl O. Igiri and Princeton N. Lyman; CSR No. 5, September 2004

*Freedom, Prosperity, and Security: The G8 Partnership with Africa: Sea Island 2004 and Beyond*
J. Brian Atwood, Robert S. Browne, and Princeton N. Lyman; CSR No. 4, May 2004

*Addressing the HIV/AIDS Pandemic: A U.S. Global AIDS Strategy for the Long Term*
Daniel M. Fox and Princeton N. Lyman; CSR No. 3, May 2004
Cosponsored with the Milbank Memorial Fund

*Challenges for a Post-Election Philippines*
Catharin E. Dalpino; CSR No. 2, May 2004
A Center for Preventive Action Report

*Stability, Security, and Sovereignty in the Republic of Georgia*
David L. Phillips; CSR No. 1, January 2004
A Center for Preventive Action Report

*Note:* Council Special Reports are available for download from CFR's website, www.cfr.org.
For more information, email publications@cfr.org.

39793678R00031

Made in the USA
Middletown, DE
25 January 2017